S0-ESC-141

Withdrawn 4596

MATTESON PUBLIC LIBRARY

801 S. School Ave.

Matteson, Illinois 60466

708-748-4431

http://www.mattesonpubliclibrary.org

Nuclear Energy

FLANGE LEAK OFF SYSTEM
TURBINE DRIVEN PUMP A
TURBINE DRIVEN PUMP B
REACTOR PRESSURE VESSEL
LEVEL
RCS WIDE RANGE
RCS MID LOOP
RCP SEALS
CHEMICAL & VOLUME CONTROL SYSTEM
REGENERATIVE HX
CHARGING
EXCESS LETDOWN HX
CCW
LETDOWN HX
LETDOWN TO PRT

DEBATING
THE ISSUES

Nuclear Energy

Marshall Cavendish
Benchmark
New York

JOHANNAH
HANEY

Copyright © 2012 Marshall Cavendish Corporation
Published by Marshall Cavendish Benchmark
An imprint of Marshall Cavendish Corporation
All rights reserved.

No part of this publication may be reproduced, stored in a retrieval system or transmitted, in any form or by any means, electronic, mechanical, photocopying, recording, or otherwise, without the prior permission of the copyright owner. Request for permission should be addressed to the Publisher, Marshall Cavendish Corporation, 99 White Plains Road, Tarrytown, NY 10591. Tel: (914) 332-8888, fax: (914) 332-1888.

Website: www.marshallcavendish.us

This publication represents the opinions and views of the author based on Johannah Haney's personal experience, knowledge, and research. The information in this book serves as a general guide only. The author and publisher have used their best efforts in preparing this book and disclaim liability rising directly and indirectly from the use and application of this book.

Other Marshall Cavendish Offices:
Marshall Cavendish International (Asia) Private Limited, 1 New Industrial Road, Singapore 536196 • Marshall Cavendish International (Thailand) Co Ltd. 253 Asoke, 12th Flr, Sukhumvit 21 Road, Klongtoey Nua, Wattana, Bangkok 10110, Thailand • Marshall Cavendish (Malaysia) Sdn Bhd, Times Subang, Lot 46, Subang Hi-Tech Industrial Park, Batu Tiga, 40000 Shah Alam, Selangor Darul Ehsan, Malaysia

Marshall Cavendish is a trademark of Times Publishing Limited

All websites were available and accurate when this book was sent to press.

Library of Congress Cataloging-in-Publication Data

Haney, Johannah.
  Nuclear energy / Johannah Haney.
     p. cm. -- (Debating the issues)
  Includes bibliographical references and index.
  ISBN 978-0-7614-4976-8 (print)---ISBN 978-1-60870-666-2 (ebook)
1. Nuclear energy--Juvenile literature. 2. Nuclear power plants--Juvenile literature. I. Title. II. Series.
 TK9148.H35 2012
 333.792'4--dc22
2010039513

Editor: Peter Mavrikis
Publisher: Michelle Bisson
Art Director: Anahid Hamparian
Series design by Sonia Chaghatzbanian

Photo research by Alison Morretta

The photographs in this book are used by permission and through the courtesy of:
Front cover: Michael Melford/Getty Images.
Alamy: Lou-Foto, 56. **Associated Press:** Dave Martin, 26; Amy Sancetta, 34; Makoto Kondo/Kyodo News, 35; Laura Rauch, 37; Jim Cole, 40; Associated Press, 40 (inset); Efrem Lukatsky, 46; Elaine Thompson, 50. **Getty Images:** Digital Vision, 1, 2-3, 4-5; Bridget Webber, 6; KEENPRESS, 8; Daniel Harris, 9; Arthur S. Aubry, 11; Dorling Kindersley, 12; Michael Melford, 14; Phil Degginger, 17; WIN-Initiative, 19; Getty Images, 21; Christopher Furlong, 29; Mandel Ngan/AFP, 30; Sarah Leen/National Geographic, 32; David Goddard, 43; Michael Dunning, 51; Michael Grecco/Hulton Archive, 52; Ted Russell, 58. **The Image Works:** Martin Benjamin, 36. **Superstock:** Science Faction, 24; Robert Harding Picture Library, 45; National Geographic, 54.
Back cover: Bridget Webber/Getty Images.

Printed in Malaysia (T)
135642

Table of Contents

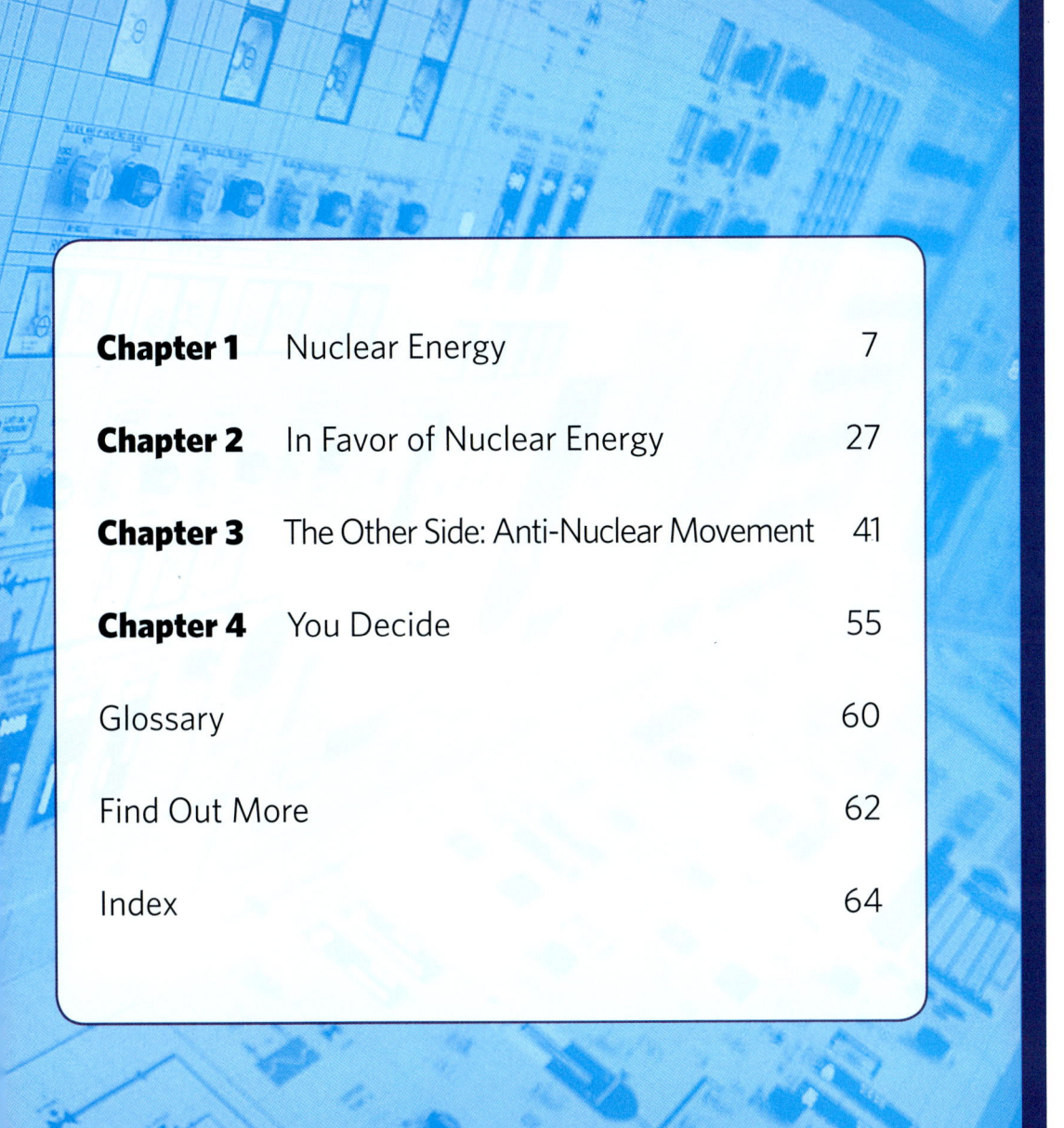

Chapter 1	Nuclear Energy	7
Chapter 2	In Favor of Nuclear Energy	27
Chapter 3	The Other Side: Anti-Nuclear Movement	41
Chapter 4	You Decide	55
Glossary		60
Find Out More		62
Index		64

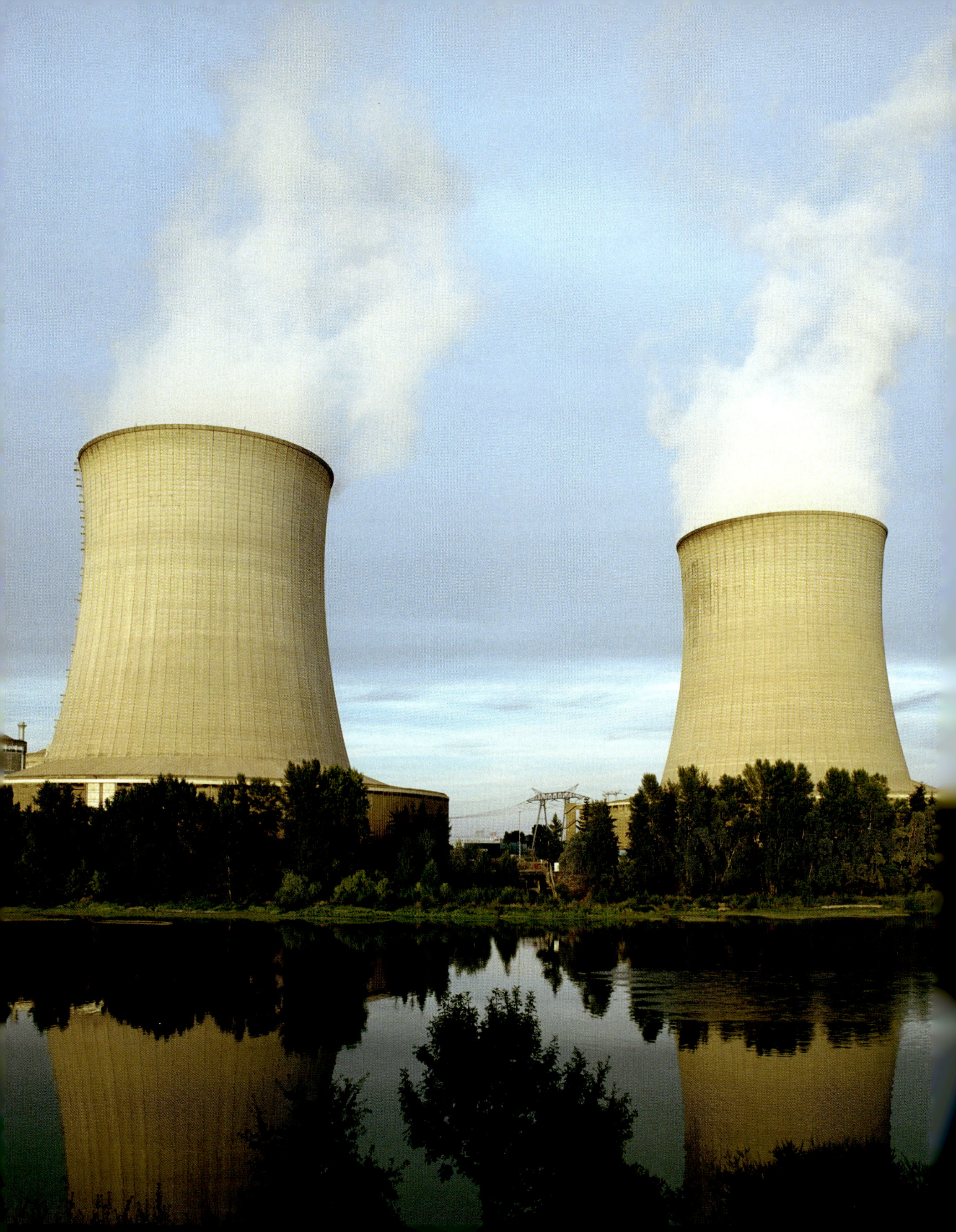

Chapter 1

NUCLEAR ENERGY

Think about all the electricity you use in a typical day from the minute you wake up and turn the lights on. An electric water heater makes your shower water a comfortable temperature. The air conditioning or heating hums along most of the day to make the air a comfortable temperature for you. Portable electronic devices such as your cell phone and your music player require a charger, which sucks up electricity when it is plugged into the wall, even if it is not connected to a device. All through the day and night, our lives run on electricity. Electricity is the life force of the modern world.

Producing electricity to keep our lives moving is not always easy. In the United States, almost two-thirds of the electricity is produced by burning **fossil fuels** such as coal, oil, and natural gas. Burning fossil fuels creates problems. It releases carbon dioxide into the environment. Chemicals and pollutants released into the air from burning fossil fuels can cause **acid rain** and **smog**. **Greenhouse gases** produced by burning fossil fuels are a contributor to **global warming**. Fossil fuels are limited in supply. No one knows for sure how long supplies of fossil fuels will last.

These twin cooling towers pour steam into the atmosphere as electricity is generated.

NUCLEAR ENERGY

GLOBAL WARMING

Global warming refers to the belief of some scientists that the temperature of the earth and its oceans is rising. They believe that major causes of global warming are the burning of fossil fuels, as well as the depletion of forests. The earth's temperature has risen about one degree Fahrenheit (about half a degree Celsius) in the past one hundred years. Scientists believe the earth's temperature will continue to rise over the next hundred years. This change in temperature over time is part of climate change. Global warming affects life on the earth in a number of ways. Melting glaciers, for example, cause the water level of oceans to rise. Even a very small change in temperature over a long period of time can cause disruption to an ecosystem. Certain types of plants and crops that have grown in a particular area for hundreds of years might not be able to grow in a warmer climate. The disruption in plant growth could affect whether animals and humans in the area are able to get enough food.

Melting ice from these glaciers runs off into the ocean.

NUCLEAR ENERGY

Many people are working on finding new ways to produce energy. Their goals include finding sustainable sources of "clean" energy—that is, sources that minimize pollution to the environment. Sustainable means that the resource is not likely to run out, as fossil fuels are. One potential source of clean, **sustainable energy** that some people think might be a solution to the world's energy problem is nuclear energy.

Nuclear energy is controversial. Some people believe it is a clean and sustainable alternative to burning fossil fuels for energy. Others believe that building nuclear reactors—the large structures that generate nuclear energy—is too expensive and that the risk of **radiation** contamination is too high. People also worry that nuclear material might fall into the wrong hands and be used to make weapons instead of energy. Before examining both viewpoints in detail, it is important to understand a little about the science behind nuclear energy.

Cooling towers help remove excess heat energy into the atmosphere.

How Nuclear Energy Works

To understand nuclear energy it is necessary to understand the building blocks of all matter—**atoms**. At the center of an atom is the **nucleus** (the plural is nuclei), a very dense area made up of particles called **protons** and **neutrons**. The number of protons in an atom determines what the atomic matter is. For example, the atoms that make up silver have forty-seven protons in each nucleus. **Uranium** is the element used in nuclear energy. Each atom of uranium contains a nucleus with ninety-two protons. Nuclear energy relies on using enriched uranium. Uranium is an element found in rock deposits. First, uranium is mined—or extracted—from rock in open-pit or underground mines. The most productive mines in the world are found in Canada, Australia, Kazakhstan, and Russia. In order for uranium to be useful in creating energy, it must first be enriched. (Enriching uranium means that the balance of isotopes is changed.) An **isotope** is an atom that has the same number of protons but a different number of neutrons. Uranium in its natural state is mostly made up of the isotope 238, known as U-238. When uranium is mined, it is made up of about 99.3 percent U-238, and just 0.7 percent U-235. These isotopes are nearly identical, but U-238 has slightly more mass. In order to use uranium as nuclear fuel, a higher proportion of U-235 is needed. The United States enriches uranium using a process called gaseous diffusion. The different isotopes are separated by mass and the final concentration of U-235 is between 3 to 5 percent. Once the uranium is enriched, it is made into uranium dioxide pellets. The pellets are packaged in long

A special suit protects this worker from possible radioactivity in the atmosphere. The instrument in his hand, called a Geiger counter, measures the amount of radioactivity.

metal tubes, called fuel rods. These rods go into the nuclear reactors for the creation of nuclear energy.

How Does Uranium Turn into Electricity?

Atoms of uranium are struck by particles called neutrons. When the neutrons strike the atoms of uranium, the atom breaks apart. This action is called **fission**.

DID YOU KNOW?
Uranium was named after the planet Uranus by German scientist Martin Klaproth who discovered the element. He originally wanted to name the element Klaprothium after himself.

NUCLEAR ENERGY

When the uranium atoms break apart, they release even more neutrons. These newly released neutrons in turn strike more atoms and thus produce more fission. In this way, a self-sustaining **chain reaction** is formed. Once the reaction starts, the chain reaction will continue over and over again on its own. In a nuclear reactor, the fission process can be controlled by rods of **graphite**. These rods, when inserted into the center of the fission material, absorb neutrons and thus slow the fission if necessary. As a result, the reaction can be controlled.

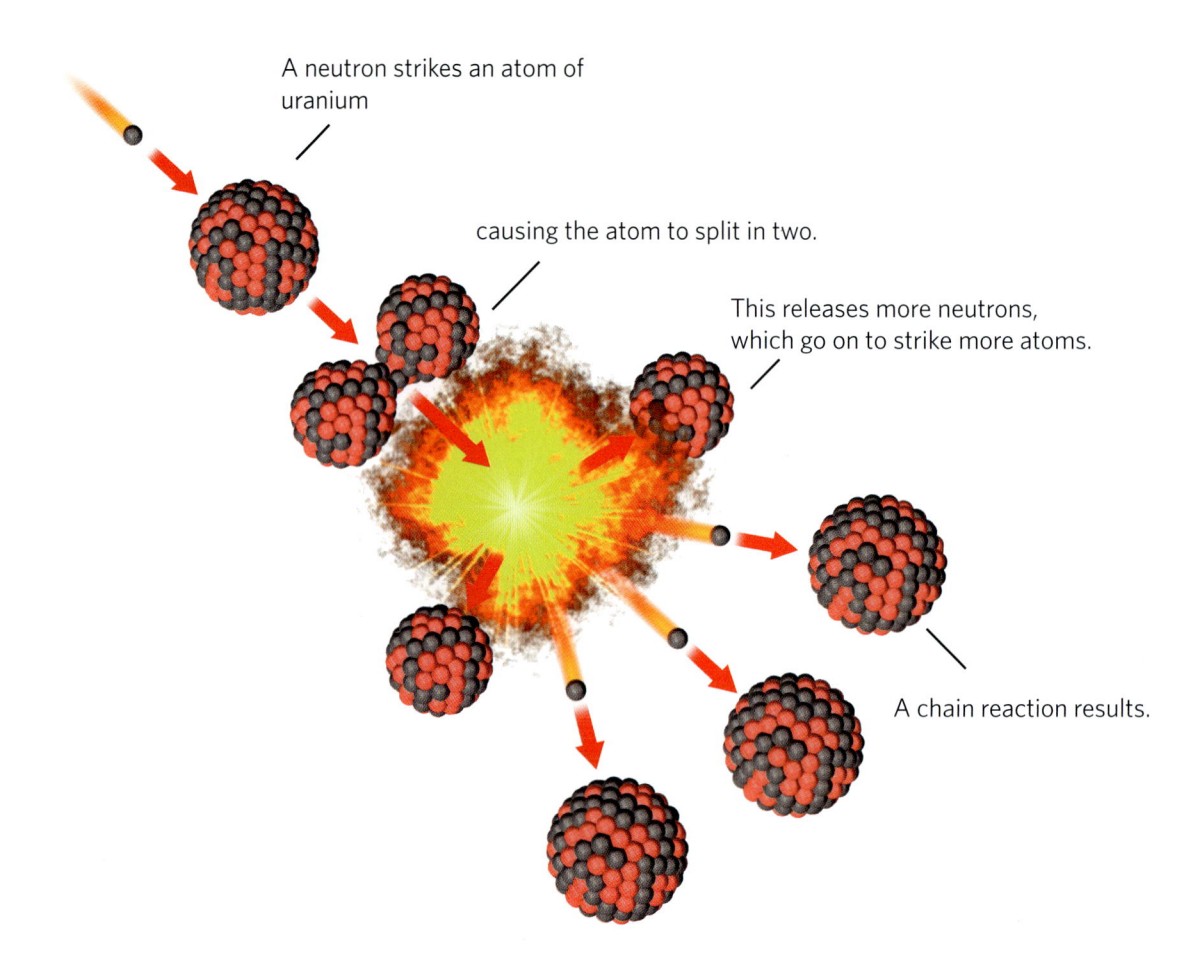

A neutron strikes an atom of uranium

causing the atom to split in two.

This releases more neutrons, which go on to strike more atoms.

A chain reaction results.

NUCLEAR ENERGY

The energy produced by fission is used to heat water. The steam that is produced from this heated water powers turbines, which extract the energy as electricity. The steam is then cooled by water, usually drawn from a body of water near the nuclear power plant. The water reduces the temperature of the steam so that it returns to liquid form. The water used to cool steam becomes hot during the cooling process. So, in many cases, the water sits in a large cooling tower so that it can cool off further. The water is then returned to its original source.

NUCLEAR POWER PLANTS

In the United States, thirty-one states have nuclear power plants. There are a total of 66 nuclear power plants and 104 nuclear reactors, and most of these sites are located in the eastern half of the country. A nuclear power plant cannot operate without a license from the Nuclear Regulatory Commission (NRC), a government organization that makes sure that the plant is safe, can operate effectively, and will not damage the environment. Each license lasts for twenty years. (Before a twenty-year renewal can be issued, the NRC must inspect the plant again.) During each inspection, NRC officials check that the building and infrastructure are well maintained and that the building is not deteriorating. They perform operating tests on the machinery and require employees to take a written test to make sure that the people operating the plant are qualified.

Radiation

The nuclear fission action produces radiation. Radiation from nuclear fission is energy that travels at high speed as particles or waves in the

NUCLEAR ENERGY

Cooling pools like this one, used to store spent fuel rods, are at least 20 feet (6 meters) deep. The nuclear reactor is connected to the storage pool by canals. In this way, the spent fuel is always held underwater and workers remain safe.

air. Different types of radiation have very different effects on humans. Radiation from nuclear reactions is called **particulate radiation**. Particulate radiation occurs in nuclear reactors as energy is being produced. It continues to be present as the nuclear fuel decays, long after the fuel is used up. This type of radiation is very dangerous. People who are exposed to very large amounts of particulate radiation can die within a few hours or days of being exposed. People exposed to particulate radiation in smaller doses are at increased risk of having children with birth defects or for developing cancer, radiation sickness, or burns.

Storing Spent Fuel

After one to two years, fuel in a nuclear reactor is removed for disposal. At this point it is called **spent fuel**. When used fuel is removed, it is still giving off radiation and heat. Over time, the amount of radiation that is given off decreases. There are two ways in which spent fuel is stored temporarily: wet storage and dry cask storage. Used fuel in **wet storage** is placed temporarily in storage ponds next to the nuclear reactor. Water in the ponds absorbs the heat the used fuel continues to give off. It also shields radiation from escaping into the air. **Dry cask storage** is another temporary storage method. After waste has cooled in wet storage for at least one year, it can be moved to a cask—a large, cylindrical steel structure. With the waste inside, the casks are welded or bolted closed. The casks are then surrounded by more steel or concrete in order to provide a leakproof barrier so that workers and

NUCLEAR ENERGY

the people living around the plant are not exposed to radiation from the waste.

In order to safely store spent fuel for long periods of time, the environment must be carefully chosen. Many people believe storing spent fuel in rock—in a mountain, for example— will help shield the environment from the spent fuel until it has decayed enough to be safe. Keeping the waste away from sources of groundwater will prevent radioactivity from contaminating water supplies. Finding a long-term solution to the storage of spent fuel is one of the main issues in the nuclear energy debate.

Nuclear Accidents

Since the first nuclear activities began in the middle of the last century, accidents have occurred with sometimes devastating effects. Two major nuclear accidents at the end of the twentieth century, at Three Mile Island, in Pennsylvania, and in Chernobyl, Ukraine (part of the Soviet Union at the time), had far-reaching effects.

THREE MILE ISLAND

The disaster at Three Mile Island began on the morning of March 28, 1979, when a pump failed to bring water to the nuclear core to cool it down. As pressure built up in the reactor core, a valve opened to release the building pressure. However, this valve did not close when it was supposed to. As a result, the coolant water drained away from the core.

Two nuclear reactors were originally built at Three Mile Island. The accident in 1979 occurred in Unit 2, pictured here on the left. Unit 2 has not operated since the accident, but Unit 1 continues to generate electricity to this day.

The core began to heat up to dangerous temperatures. The instruments that helped the technicians monitor conditions were confusing, and operators made the problem worse by turning off the emergency water system. Two hours and forty-five minutes after the first pump failure that set events in motion, radiation alarms began to sound, and a state of emergency was declared.

DID YOU KNOW?

Just twelve days before the accident at Three Mile Island, a movie called *The China Syndrome*, starring Jane Fonda, Jack Lemmon, and Michael Douglas, was released in the United States. The movie was about the cover-up of a nuclear accident. The movie fueled public panic about the Three Mile Island accident.

NUCLEAR ENERGY

In the end, more than half of the nuclear core was exposed. Officials immediately recommended that pregnant women and small children evacuate the area because of their increased risk for damage from radiation exposure. The United States Regulatory Commission (NRC) reported that the amount of radiation that the average person in the area was exposed to after the accident was about 1 millirem (a chest X-ray would expose a person to about 6 millirems). This level of radiation is not dangerous to most people. However, some scientists argue that much more radiation was released.

The disaster at Three Mile Island raised the issue of the safety of nuclear energy in the United States. It highlighted the need for evacuation plans for communities with nuclear power plants. It initiated recommendations for better safety measures for nuclear power production. It also caused many people to wonder whether the risks involved in generating electricity from nuclear power were worth the benefits. As a result of Three Mile Island, the future of nuclear energy in the United States was put into more serious question. Just seven years later, another accident—this time thousands of miles away—made the whole world seriously consider the safety of nuclear energy.

CHERNOBYL

On April 26, 1986, there was a terrible accident at the nuclear reactor in Chernobyl, Ukraine. Before the accident, scientists were experimenting with a new way to handle power outages at the plant. As part of the

Abandoned buildings like these are a common sight in Prypiat, the town near the Chernobyl Power Plant. Residents left everything they had behind when they evacuated following the devastating nuclear accident.

test, they disabled many of the automatic safety features in the reactor. (Remember that to generate nuclear energy, a chain reaction of fission is performed in a controlled way.) In the Chernobyl catastrophe, the chain reaction was not being controlled. The core began to melt down. As temperatures rose higher and higher, a steam explosion ripped the cover off the reactor. A second explosion sent fuel and burning graphite flying. This debris started fires in nearby buildings.

DID YOU KNOW?
Today, the area around Chernobyl has become a haven for wildlife. With no human interference, some species that had almost disappeared have returned and are thriving in the seclusion.

JAPAN'S FUKUSHIMA DAIICHI DISASTER

On March 11, 2011, a magnitude 9.0 earthquake hit near the coast of Honshu, Japan, causing severe damage to buildings and triggering a massive tsunami. In the wake of the destruction sat the six reactors at the Daiichi nuclear plant in Fukushima. Loss of power and failure of backup systems compromised the cooling systems at the plant. In an effort to replenish water in spent fuel pools, authorities tried spraying water into the buildings and dropping tons of water from helicopters. The full effects of the nuclear disaster in Fukushima and elsewhere will not be fully known for some time.

An immense amount of radioactive debris was released into the atmosphere. The lighter particles were carried by wind to the surrounding nations Belarus and Russia and in smaller amounts throughout Europe. Pripyat, the town closest to the Chernobyl nuclear reactor, was not evacuated right away. When it was evacuated about a day after the accident, residents were told they would be gone for only a few days. In reality, most never returned because the radioactive dust buried all their personal belongings, and it was not safe to go home.

The disaster at Chernobyl claimed lives and continues to affect people's health to this day. The initial explosions killed two workers. Several firefighters who came to fight the ten-day graphite fires died of radiation exposure within several weeks of the accident. The United Nations estimates that fifty-six people died as a result of exposure to radiation from the accident. The International Atomic Energy Agency estimates that the death toll will reach 4,000 when considering people who are at risk for developing cancer from the high levels of radiation. Anti-nuclear activists say this number is too low.

The town of Pripyat is still abandoned. The 18.5-mile (30 km) zone around the plant, called the "exclusion zone," will not be safe for hundreds of years to come.

Nuclear Energy Policy in the United States

Ever since the United States first developed nuclear energy, it has been establishing laws and regulations to help manage how nuclear energy is used in the United States. The first piece of nuclear energy legislation, the Atomic Energy Act of 1946, set up the Atomic Energy Commision (AEC) to oversee the nuclear industry. Previously, the military held this responsibility. The AEC is made up of five civilians, appointed by the president of the United States. Their responsibilities include ensuring the safety and development of nuclear weapons as well as of nuclear energy. The Atomic Energy Act of 1954 made a few changes to the original law. It said that any private company setting up a nuclear facility must obtain a license from the Atomic Energy Commission. The AEC was given the additional responsibility of establishing and enforcing rules and regulations for the health and safety of power plant workers, the environment, and the general public.

The Energy Reorganization Act of 1974 was pivotal in establishing the nuclear oversight in existence

Nuclear regulators provide oversight to make sure nuclear power plants are being run and managed safely and in accordance with laws and regulations.

Nuclear Energy Timeline

1789 – German chemist Martin Klaproth discovers uranium.

1896 – Uranium's radioactive properties were discovered by French scientist Henri Becquerel.

1930s – Nuclear fission is discovered through contributions of many scientists working at this time, including Enrico Fermi, Otto Hahn, Fritz Strassman, Lise Meitner, and Otto Frisch.

1946 – The Atomic Energy Act of 1946 is signed into law, creating the U.S. Atomic Energy Commission to oversee the nuclear energy activity of the United States.

1951 – A nuclear reactor in Idaho, the EBR-1, generates electricity from nuclear energy for the first time, producing enough power to light four lightbulbs.

1954 – The Obninsk Nuclear Plant in the former USSR is the first nuclear plant to generate electricity for a power grid.

1955 – The EBR-1 in Idaho has a partial core meltdown.

1956 – The first commercial nuclear power plant opened in England.

1965 – The first nuclear reactor operates in space.

1974 – The Nuclear Regulatory Commission (NRC) becomes the regulatory body of nuclear energy in the United States.

1976 – One of the largest protests against nuclear power development was staged in Spain, with more than 200,000 protesters.

1978 – Austria votes to ban nuclear power.

March 28, 1979 – A failure of the cooling system led to a partial core meltdown in Pennsylvania, known as the Disaster of Three Mile Island.

April 26, 1986 – The Chernobyl nuclear power plant in the Ukraine becomes the worst nuclear accident to date.

1993 – 109 nuclear power plants operating in the United States provide one-fifth of the country's electricity.

2002 – The Nuclear Power 2010 Program strives to find the right places to build new nuclear power plants in the U.S.

2005 – The Energy Policy Act is signed into law by President George W. Bush. This included tax breaks and other financial incentives in support of developing nuclear and other alternative energy sources.

March 11, 2011 – A magnitude 9.0 earthquake and resulting tsunami damage the nuclear reactors in Fukushima, Japan, causing failure of cooling systems and leading to the threat of nuclear meltdown.

today. This act abolished the Atomic Energy Commission. Instead, a new agency, the Nuclear Regulatory Commission (NRC), was set up to regulate the safety of nuclear energy production. The Department of Energy was given the responsibility of overseeing nuclear weapons development and safety.

The Nuclear Non-Proliferation Act of 1978 was an important law in helping safeguard against anyone using nuclear material for weapons. It set new rules for exporting nuclear material to other countries.

In the early 1980s, two new laws were passed that specifically targeted spent fuel. The Nuclear Waste Policy Act of 1982 stated that the government would be responsible for providing a safe place to store spent fuel over the long term. The Low-Level Radioactive Waste Policy Amendments Act of 1985 requires each individual state to ensure that spent fuel created within its borders is stored in the short term.

The Price-Anderson Nuclear Industries Indemnity Act, signed in the early 1950s, limited the amount of money a nuclear facility would

Mid-level radioactive waste includes equipment and supplies that come into close contact with fuel rods. It is sealed into concrete containers like this one for disposal.

have to pay if there was an accident. This act was extended in the Energy Policy Act of 2005. The 2005 act also established several laws to help the nuclear energy industry. Nuclear reactors, for example, are granted a tax credit the first eight years of operation. Also, loans that are taken out to build new nuclear reactors are guaranteed by the government. These provisions are important because they make it less risky for investors to build new nuclear plants.

There are many things to consider when deciding whether nuclear energy is a good idea or not. In a time of global warming, dwindling nonrenewable resources, and political tensions caused by trade of oil, finding alternative methods of creating energy is an issue that is becoming more important. Is nuclear energy the right answer?

WHAT DO YOU THINK?

What financial costs and risks do you think go into building and maintaining a nuclear power plant?

Are the events at the Chernobyl plant and Three Mile Island relevant to the current nuclear energy debate? Why or why not?

How important is the issue of storing spent fuel to the nuclear energy debate?

Do you think the Price-Anderson Nuclear Industries Indemnity Act is a good way for the United States to encourage nuclear energy use?

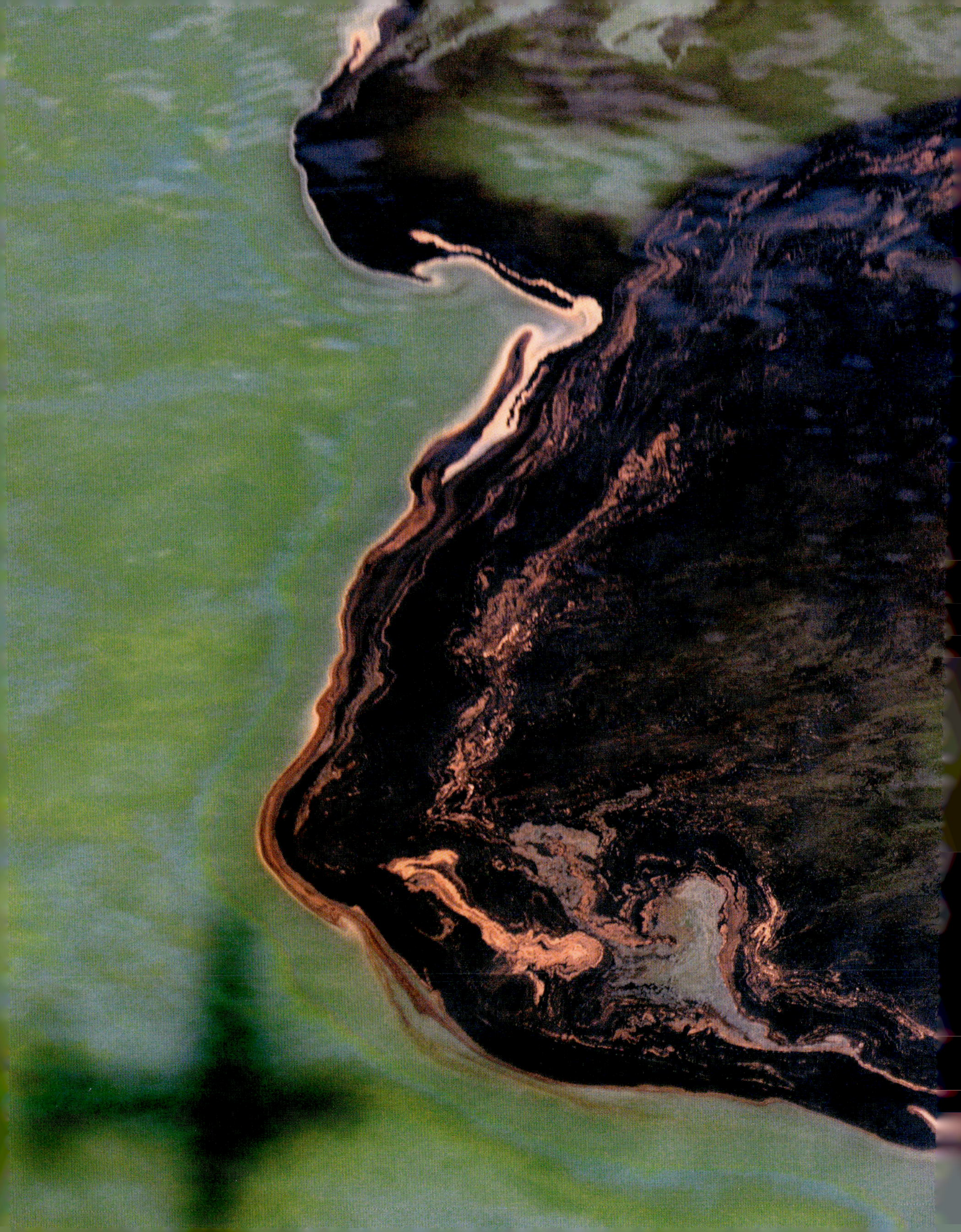

Chapter 2

IN FAVOR OF NUCLEAR ENERGY

April 2010 was difficult for the coal and oil industries. In April, the Upper Big Branch mine near Montcoal, West Virginia, suffered an explosion that killed twenty-nine coal miners. It was the worst coal mining accident in the United States in almost forty years. Just fifteen days later, the Deepwater Horizon oil rig in the Gulf of Mexico exploded, killing eleven workers and injuring seventeen. After the explosion, the hundreds of thousands of barrels of oil that gushed into the waters of the Gulf of Mexico caused a great deal of damage to the environment. The effect on aquatic and onshore ecosystems and on the fishing and tourism industries is yet to be determined. Both disasters caused deep anguish among Americans from coast to coast and highlighted the danger and risk involved in obtaining fossil fuels. As already stated, supplies of fossil fuels are limited and at some point may run out. The solution may lie in finding alternative sources of energy.

Few people dispute the fact that the current methods of generating most of the electricity in the United States is not sustainable. As supplies of the earth's resources dwindle, some people think that nuclear energy is one solution to these challenges.

When the Deepwater Horizon oil rig exploded in the summer of 2010, oil formed an immense slick on the Gulf of Mexico. The environmental impact may not be fully understood for many years to come.

NUCLEAR ENERGY

OVERVIEW OF THE PRO-NUCLEAR ENERGY STANCE

SAFETY
- Although accidents are possible, the current nuclear technology makes it very unlikely that accidents like those at Chernobyl or Three Mile Island will occur again.
- The quantity of spent fuel is small enough that it can be safely stored on-site at nuclear plants, using wet storage or dry cask storage.
- The amount of radiation that a nuclear power plant gives off is low enough to be safe for people living and working near the plants.

ENVIRONMENT
- Burning fossil fuels has immediate and known consequences to the environment. Nuclear energy is much cleaner, and the benefits outweigh the risks.
- The supply of uranium is abundant, and very little is needed to produce a large amount of energy. In addition, new stores of uranium are likely to be discovered, meaning the supplies could last for centuries.

COSTS
- Nuclear-generated electricity is cost-effective.
- Financial advantages, such as loan guarantees, make it less risky for new nuclear facilities to be built.

This wind turbine generates clean electricity right next to coal-burning power plants. Although wind energy uses a renewable resource, it cannot create energy in the same capacity as coal or nuclear energy.

Proponents of nuclear energy often compare the environmental impact of nuclear energy with that of power plants that burn fossil fuel. The United States gets a little more than 70 percent of its electricity from coal and natural gas. Coal energy is produced by burning coal to heat water, a process that creates steam. The steam turns a turbine, and electricity is generated. Burning coal to produce the steam, however, is a dirty process; it produces emissions of carbon dioxide, sulfur dioxide, nitrogen oxides, and mercury compounds. These emissions are contributing greatly to global warming. Natural gas produces similar emissions, although in smaller amounts. Nuclear energy does not pose this problem. The emissions from the nuclear energy process are practically zero. In fact, no emissions are created from the plant itself from generating electricity. Some natural resources, such as oil

Mountaintop mining/valley fill is a process in which the top of a mountain is demolished in order to expose available coal. The coal is mined from the mountain, and the rock debris is moved to a nearby valley.

and gas, are used in the process of mining and milling uranium, but compared with the quantities of resources used in coal and natural gas plants, the amount is minimal.

Both coal and natural gas, like oil, are nonrenewable resources. They also cannot be replenished. Coal and natural gas took millions of years to develop from organic materials that have been exposed to extreme heat from within the earth. The earth has a limited supply of these natural resources. For example, it is estimated that there are only 268 billion tons (243 metric tons) of coal available in the United States. In 2003 the U.S. used about 1.1 billion tons (997 million metric tons) of coal for energy. Uranium, which, as mentioned earlier, is needed for the production of nuclear energy, is also a nonrenewable resource. Right now the United States uses 77,000 tons (70,000 metric tons) of uranium each year. At this rate, the uranium that has been discovered so far would last about 230 years. However, it is very likely that more uranium will be discovered. In addition, nuclear energy generates electricity very efficiently. A large quantity of energy can be taken from a relatively small amount of uranium. With the development of better nuclear fuel technologies, the amount of uranium that is necessary to produce that same quantity of electricity might be reduced, and it could be that less than half the amount of uranium will be required in the future. Taking these factors into account, many people believe that uranium will be a far longer-lasting resource than coal or natural gas.

Both coal and nuclear energy produce solid wastes. However, proponents of nuclear power point out that the volume of waste produced by nuclear power plants is far less than that of coal-powered plants. About 2,200 tons (2,000 metric tons) of waste is produced per year by the U.S. nuclear power plants. A single coal plant produces about 125,000 (113,000 metric tons) tons of ash per year and 193,000 tons (175 metric tons) of sludge per year. The ash and sludge produced in a coal plant is often placed in landfills. Toxic substances in the waste, including arsenic, mercury, chromium, and cadmium, can potentially contaminate soil and drinking water. Spent fuel from nuclear energy production is certainly capable of damaging the environment as well; however the volume of waste is far less, and each plant stores the waste carefully on-site.

NUCLEAR ENERGY

DID YOU KNOW?
If someone stacked all the spent fuel in the world from the last forty years, it would fill a football field in layers 21 feet (6.4 meters) deep.

Why Not Stick to Renewable Energy Sources?

Renewable energy sources—such as wind and solar power—are cleaner than coal or natural gas, and they carry less risk than nuclear power. Some people suggest that renewable sources should be used instead of coal, natural gas, or nuclear energy. Although these methods are green—that is, environmentally friendly—proponents of nuclear power believe that a mix of sources is needed. At their current level of technology, wind power, solar power, hydroelectric power, and other renewable power sources cannot provide enough of a base of electricity to power modern life. Nuclear power, however, can generate enough electricity to power big cities and smaller towns alike.

Voters in Sacramento, California decided to shut down the Rancho Seco nuclear power plant in 1989. Since then, the site has become home to an array of solar panels that provide some of the energy for the community.

Risks of Nuclear Energy

People who support the use of nuclear energy do not deny that there is some threat of accidents, radiation exposure, nuclear proliferation, and other negative effects from its use. However, many believe the threat is not as high as anti-nuclear activists claim. People in favor of nuclear power also say that since it is a fact that burning fossil fuels is damaging the environment now, it is better to look to nuclear power as a better alternative. After all, in most cases nuclear energy can be generated safely and with less damage to the environment.

Accidents

The risk of a nuclear accident is a serious part of the debate and not taken lightly by people on either side. The accident at Three Mile Island, because it happened in the United States, highlighted that risk for Americans. The Chernobyl disaster is a tragedy not to be forgotten. (The events at Fukushima, Japan, are still being analyzed.)

However, people in favor of nuclear energy believe that such accidents would not happen today. The technologies used at the Chernobyl site in 1989 are vastly different from standards used today. The safety features in place nowadays at nuclear energy facilities would make another accident like Chernobyl unlikely. Although the accident at Three Mile Island was frightening for people in the community and across the country, it demonstrated that the technology in a nuclear power plant works to prevent a series of errors from becoming a major nuclear event. Three Mile Island suffered a partial

On June 24, 1998, a tornado hit the community of Oak Harbor, Ohio—and its nuclear power plant. In this photo, workers are monitoring the reactor from the control room after the tornado. Despite the storm, no radiation escaped and no one was hurt at the power plant.

core meltdown, yet no one was hurt in the incident. In fact, no one has died in the United States from a radiation accident. About five thousand people die each year in coal-mining accidents, however.

DID YOU KNOW?
Nuclear power plants operating today give off minuscule amounts of background radiation to workers and neighboring communities. A person would be exposed to many times more radiation from a chest X-ray.

Spent Fuel

Used fuel can be recycled so that any reusable remnant can be made into new fuel rods. After the waste is held in temporary storage, it can be permanently stored. As of 2010 there were no permanent waste storage sites anywhere in the world. However nuclear energy proponents are not too concerned about the situation. The volume of

IN FAVOR OF NUCLEAR ENERGY

spent fuel is still small, and proponents argue that there is time to find the right long-term storage solutions. In the meantime, there is plenty of space for short-term storage at the nuclear power plants. In fact, there is an advantage to keeping spent fuel in short-term storage since the longer it is held there, the safer it is to handle, as the levels of radioactivity continue to decrease with time.

Nuclear Proliferation and Terrorism

Nuclear proliferation and the threat of U.S. nuclear reactors being the target of a terrorist attack are unfortunate concerns in the political atmosphere of the twenty-first century. Most pro-nuclear activists argue that proliferation is a real threat. However humankind cannot ban every useful technology that could be used for evil by some people. If we did, we would have to ban kitchen knives, airplanes, and thousands of other inventions.

The threat of nuclear power plants being terrorist targets is another real concern. After the events of September 11, 2001, in

A cylinder containing nuclear waste is loaded onto a truck to be shipped to a storage facility.

which terrorists flew hijacked airplanes into buildings in New York City and Washington, D.C., people have grown concerned about the effect a similar attack would have on a nuclear reactor. People in favor of nuclear energy do not deny that possibility but point to the reinforcements of modern nuclear reactors, which have a concrete containment structure to protect against such attacks. The fact that the used-fuel storage is usually located below ground level offers another layer of protection. Pro-nuclear activists also contend that chemical plants and high-profile political landmarks are more likely to be targeted.

A Future of Nuclear Energy

Proponents of nuclear energy envision a future in which nuclear energy will take the place of dirtier coal and gas energy. Advancing nuclear technologies would make producing energy safer than it has ever been in the past and eliminate the risks associated with obtaining fossil fuels for energy. People in favor of nuclear energy believe that decreasing the reliance on foreign oil will ease tensions between the United States

This woman openly displays her support for nuclear power.

YUCCA MOUNTAIN REPOSITORY

The Yucca mountain range, located 80 miles (130 kilometers) northwest of Las Vegas, was supposed to become the first long-term storage facility for spent fuel. The rock that makes up the Yucca Mountain, called tuff, is composed of volcanic ash, which is believed to be a good host for nuclear waste. However in 2009, the U.S. Department of Energy tried to block the project. Instead, the department has stated it will work on finding a different way to dispose of spent fuel materials. But judges from the Nuclear Regulatory Commission say that because the plan was part of the Nuclear Waste Policy Act of 1982, the Department of Energy cannot eliminate the plan without separate legislation. That would mean that politicians would have to vote whether to stop preparing Yucca Mountain for nuclear waste. Such legislation could take several years to enact. For now, the future of the Yucca Mountain Repository is uncertain. It will be one of the many aspects of the nuclear energy debate.

Workers at the tunnel entrance of the Yucca Mountain Project.

and oil-rich nations. Their goal is to see an expansion in the percentage of electricity derived from nuclear energy both in the United States and abroad.

ENVIRONMENTALISTS SUPPORTING NUCLEAR ENERGY

The biggest opponents of nuclear energy have traditionally been environmentalist groups. However, some of these groups are beginning to give nuclear energy a second look. Patrick Moore, one of the cofounders of Greenpeace, a major environmentalist group founded in 1971, changed his mind about nuclear energy and now believes it is an important part of a sustainable energy future. Environmentalists for Nuclear Energy, another eco-activist group, believes nuclear energy is clean. Its members contend that the construction materials used in nuclear energy production, compared with those used in wind and solar power production, make it more efficient, that nuclear energy produces very little waste, and that the waste it does produce is contained and does not contribute to the greenhouse effect.

WHAT DO YOU THINK?

Proponents of nuclear energy, focusing on the future, anticipate that yet-to-be-developed technology will make the earth's uranium supply last longer. Do you think that is a fair and solid argument? Why or why not?

Does the accident at Chernobyl prove that nuclear energy is unsafe? Why or why not?

Are you comfortable with the idea that there is no concrete plan for storing spent fuel over the long term? Why or why not?

Do you think the threat of nuclear proliferation is serious enough to outweigh potential benefits of nuclear energy?

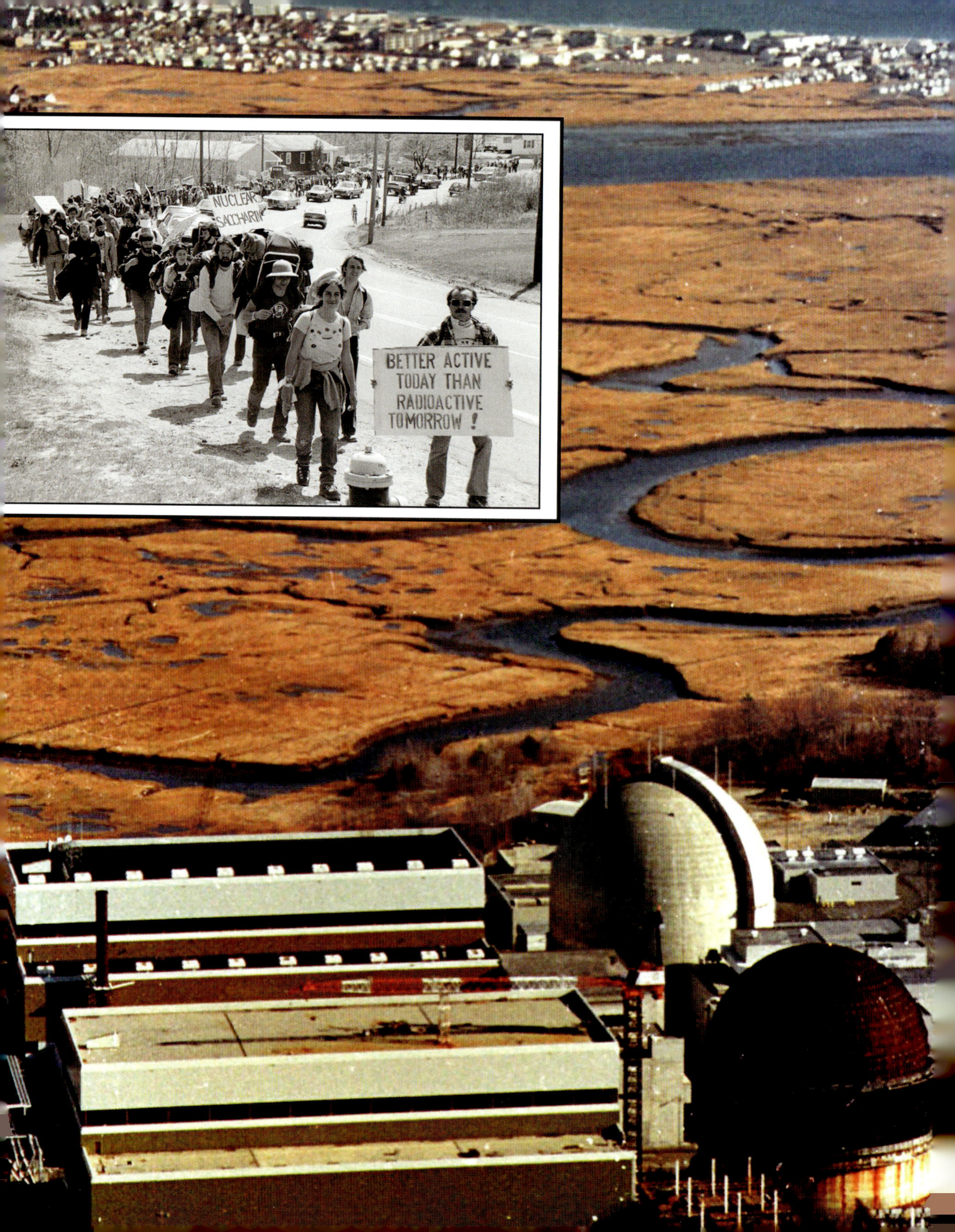
NUCLEAR
BETTER ACTIVE TODAY THAN RADIOACTIVE TOMORROW !

Chapter 3

THE OTHER SIDE: ANTI-NUCLEAR MOVEMENT

In 1977, in the small town of Seabrook, New Hampshire, two thousand protesters who were part of an anti-nuclear activist group called the Clamshell Alliance arrived at the site of a proposed nuclear power plant. They set up tents and prepared for a long stay at the site. When police arrived, they did not resist arrest. More than 1,400 people were taken into police custody for trespassing. They could have been released from jail by posting $500 bail, but almost all of them refused. Thus, the state of New Hampshire had the expense of paying for the care of the protesters while they were waiting to be heard by a judge. Their goal was to stop the construction of the plant. One of the protesters said, "We feel Seabrook in particular and nuclear power plants in general are life and death issues. We are acting in self-defense." Partly owing to the costs associated with the Clamshell Alliance protest, the builders of the Seabrook nuclear power plant had financial difficulty. In 1990, eighteen years after the plans were first submitted, the nuclear power plant finally began to generate electricity for customers. The original plans for the Seabrook Station included two nuclear reactors, but because of financial issues, the second reactor was never built.

The Seabrook plant is pictured here, with an image from one of the rallies that was held to protest its construction.

NUCLEAR ENERGY

OVERVIEW OF THE ANTI-NUCLEAR ENERGY STANCE

SAFETY
- The accidents at Three Mile Island, Chernobyl, and Fukushima Daiichi in Japan, show that nuclear energy is too risky.
- Without established solutions for long-term storage, spent fuel should not be generated.
- Scientists may be underestimating the amount of background radiation that is being emitted from nuclear power plants.

ENVIRONMENT
- The impact of a nuclear accident—or a nuclear weapon derived from material intended for production of energy—would have an immediate catastrophic effect on the environment.
- Uranium, like coal, oil, and natural gas, is a nonrenewable resource.

COSTS
- It costs too much for nuclear plants to be built, operated, and decommissioned.

THE OTHER SIDE: ANTI-NUCLEAR MOVEMENT

THE ENVIRONMENT

There are many concerns about how nuclear power production will affect the environment. The concerns begin with the mining of the uranium that is necessary for the fission process and include nuclear-waste management, thermal pollution, and the risk of accidents. Anti-nuclear activists argue that the environmental impact is too great.

A number of concerns arise with the mining of uranium. To begin with, the amount of uranium in rock is very small, and so a lot of rock must be mined in order to get a significant amount of uranium. The leftover rock contains **radon gas**, which is known to be

DID YOU KNOW?
One alternative energy source that is more environmentally friendly than nuclear energy is known as clean-coal technology. This technology is being developed so that coal can continue to be used to produce electricity, but with fewer damaging effects on the environment.

One environmental concern of nuclear energy is thermal pollution to bodies of water near nuclear plants. Here, a water outlet discharges cooling water from a nuclear plant in England. When water is discharged at a higher temperature, it can have a dramatic effect on the ecosystems of the waterways.

NUCLEAR ENERGY

a cancer-causing agent. **Waste rock**, as the leftover rock is known, also contains trace amounts of radioactive elements. However, the U.S. government does not consider this rock to be radioactive waste, so there are no regulations or safety requirements for its disposal. Often waste rock is stored in piles on or near the site of the mine. People are concerned that radon gas and radioactive particles are contaminating the air and groundwater.

> **DID YOU KNOW?**
> The Radiation Exposure Compensation Act was passed in 1990 in order to compensate uranium mine workers, mill workers, ore transporters, and others exposed to nuclear testing during the Cold War. People who developed certain forms of cancer from exposure to radon gas are eligible to receive up to $100,000 in compensation.

Once nuclear fuel is at the nuclear power plant and is being used to generate electricity, a new kind of waste is produced along with the electricity: waste heat. Water is cycled through the power plant in order to cool the nuclear reactor core. Heated water that is discharged from nuclear power plants, like the water discharged from coal-burning plants, can have a devastating impact on local streams, lakes, and rivers. Many aquatic species of fish and plants are very sensitive to changes in temperature.

Opponents of nuclear energy are concerned about the effect of spent fuel on the environment. As of 2010, there were no permanent waste storage facilities anywhere in the world. One major concern is the energy it would take to move spent fuel into a long-term storage facility once one becomes available. For example, when the plan for the nuclear storage facility at Yucca Mountain in Nevada was

Those opposed to nuclear energy are concerned that yellowcake, the concentrated form of uranium seen here, is dangerous to nearby communities and could be stolen and used to create dirty bombs.

under way, opponents pointed out that bringing spent fuel there from all over the country would come at a great expense in terms of carbon dioxide output from trucks or trains. They also warned that transporting spent fuel close to homes in the United States would expose people along the route to danger if something went wrong. The same concern is held when concentrated uranium that has been mined and milled, known at this stage as **yellowcake**, is shipped.

The amount of time it takes for spent fuel to become safe is estimated to be ten thousand years or more. It is difficult to predict conditions in a potential storage site for that long. Critics worry that natural events such as earthquakes could disrupt waste sites and cause radioactive waste to be released into the environment. Others are concerned that over such a long period of time, conditions in a waste site could change drastically. For example, if the place where waste is stored is

NUCLEAR ENERGY

very dry at the time of storage, it is possible that over time the climate could change, and the area could become wet. Such a change could put water supplies at risk.

> **DID YOU KNOW?**
> The Price-Anderson Act, which limits the amount of money nuclear reactor operators would have to pay if something goes wrong, is challenged by many anti-nuclear activists. They say the law would cost taxpayers too much money in case of an accident.

Threat of Nuclear Accidents

Perhaps one of the biggest concerns related to nuclear energy production is the effect a potential accident would have. The effects from the Chernobyl accident on the environment show that nuclear accidents have grave environmental consequences. In the Wormwood Forest, an area near the Chernobyl site, all the pine trees turned a reddish brown color after the massive radiation release. For this reason it became known as the Red Forest. The pine trees were bulldozed and buried in the soil and topped with a thick layer of sand. Today the new trees growing in their place have significant **mutations**. Studies have shown that birds living in the area around Chernobyl have a slower

New pine trees growing in the forests surrounding the Chernobyl nuclear power plant demonstrate evidence of mutations, such as these trees with browned tips.

THE OTHER SIDE: ANTI-NUCLEAR MOVEMENT

reproductive rate and a lower survival rate. Experts are worried that animals with gene mutations from radiation will pass those mutations on to animals in the general population.

WHAT IS NUCLEAR FALLOUT?

Nuclear fallout is the leftover radiation that "falls out" of the sky in the form of dust and debris after a nuclear explosion. One of the single most devastating cases of nuclear fallout occurred after the atomic blasts in Hiroshima and Nagasaki, Japan, in World War II. The United States dropped the two atomic bombs in August 1945, and the immediate results from the blast were devastating. Over 70,000 people were killed in Hiroshima and 40,000 in Nagasaki. Because of the lingering effects of nuclear fallout, the death toll five years after the attack was estimated at around 200,000 in Hiroshima and 140,000 in Nagasaki. These numbers include deaths from cancer and birth defects and other radioactive illnesses from the fallout.

The impact of radiation on human health is of great concern to anti-nuclear activists. All nuclear power plants give off a certain amount of radiation through the normal course of operations. The amount of radiation that can be legally released, the "permissible" dose, is tightly regulated. However, anti-nuclear activists argue that a permissible dose does not mean that it is a safe dose. They also argue that the standards for measuring and reporting the amount of radiation released are flawed and that more radiation is being released than is known or reported.

Safety

Safety is a major concern for opponents of nuclear energy. One aspect of nuclear energy safety is the levels of radiation that workers at nuclear power plants are exposed to, as well as the levels people in and around the communities where nuclear power plants are located are exposed to. As discussed, anti-nuclear activists believe no level of radiation exposure is safe and that the actual amount of radiation being released is higher than reported.

Another dimension to nuclear energy safety is the risk of an accident at a nuclear power plant. There are other effects as well. Anti-nuclear organizations claim that the financial costs of cleaning up a nuclear accident would be paid for with tax money instead of by the nuclear industry. They believe that because owners of nuclear power plants would not be responsible for environmental cleanup in the event of an accident, they have less incentive to spend the money needed to keep the reactor safe and accident-free. Another concern is that there are no well-developed plans to evacuate an area in the case of a nuclear accident. Having a well-defined evacuation plan if necessary or a plan to inform a community of hazardous conditions is seen as critical to avoiding widespread exposure to radiation after a nuclear accident.

Terrorism

Nuclear power plants are potential targets for terrorists both in the United States and throughout the world. One concern is the targeting

THE OTHER SIDE: ANTI-NUCLEAR MOVEMENT

MUTATIONS

> Exposure to radiation in high amounts or for long periods of time can cause mutations. A mutation is a permanent change to DNA. Mutations that occur from radiation are known as acquired—or somatic—mutations. The human body can repair the damage from small doses of radiation—for example, the radiation a person receives during a medical X-ray. When the dose of radiation is high, the body cannot repair this damage, and mutations may occur. Humans, animals, plants, and insects can all experience genetic mutations as a result of exposure to radiation. The types of mutations vary widely. In fact, the mutations caused by radiation are no different from mutations that occur randomly in nature. However, exposure to radiation makes these mutations occur more frequently than mutations in nature.

of the radioactive wastes that are held in pools or in dry storage at the site of nuclear plants. Another would be an attack on the reactors themselves or sabotage of the water system in order to cause a core meltdown. According to a Congressional Budget Office paper, "The human, environmental, and economic costs from a successful attack on the nuclear power industry could far exceed the value of the nuclear plants themselves." People who oppose nuclear energy believe that the risk of a terrorist attack on a nuclear facility is not worth taking.

Another concern is that terrorists could somehow steal or purchase on the black market spent-fuel materials. This nuclear material is most often used to make radiological weapons, commonly known as "dirty bombs." The radiation from the explosion of spent-fuel material would cause significant localized damage and fallout.

Nuclear facilities have security measures in place to prevent a

This simulation shows the magnitude and power that a dirty bomb could have.

thief or terrorist from getting into a plant in order to steal radioactive material. However, some are concerned that a terrorist or someone who wants to sell nuclear material on the black market could get access by getting a legitimate job in a nuclear plant.

Financial Costs of Nuclear Energy

The money needed to build, operate, and decommission—that is, take out of service—a nuclear facility is a major concern in the anti-nuclear-energy debate. The high costs of building new plants, disposing of high-level waste (the waste that has the most radioactivity), and taking older plants out of service would be better diverted to other forms of energy generation.

THE OTHER SIDE: ANTI-NUCLEAR MOVEMENT

NUCLEAR WEAPONS

Often when people think of nuclear energy, they also think of nuclear weapons. Nuclear weapons also use fission in order to generate energy. However, the fission is not controlled, as it is when nuclear power is generated, and so the result is an explosion. Nuclear bombs are very powerful. Nuclear proliferation is a term people often use to describe the spread of nuclear weaponry and warfare throughout the world.

One source of material for nuclear weapons is a by-product of the generation of nuclear energy. When spent fuel is reprocessed, a substance called separated plutonium is produced. It is readily usable in nuclear weapons. There are more than 297 tons (270 metric tons) of separated plutonium in the world. Although nuclear reactors have heavy security, a major concern is that nuclear material will fall into the wrong hands.

How Does the Anti-Nuclear Movement Work toward Ending Nuclear Energy?

Anti-nuclear groups have a variety of ways to promote their message. One way is to try to influence government policy, for example, by working with lawmakers to try to increase the restrictions put on nuclear power plants. A large part of their work, however, is to educate the public about why they oppose nuclear power and to try to persuade others

Nuclear weapons like the missile shown here could have a devastating effect on world diplomacy and warfare.

Musicians like Bruce Springsteen have been active in the anti-nuclear movement, hosting events and concerts to raise awareness about issues related to nuclear energy.

to join the movement and work against nuclear energy. Other ways that anti-nuclear groups get their message to the public and the nuclear industry include distributing pamphlets and fact sheets, hosting concerts to benefit their cause, and peaceful protesting, usually in large groups.

A Nuclear-Free Era of Renewable Energy

People who oppose nuclear energy contend that reliance on fossil fuels is unwise because they will eventually run out. However, they argue

THE OTHER SIDE: ANTI-NUCLEAR MOVEMENT

that nuclear energy also has significant environmental disadvantages. Uranium might seem more abundant than coal, oil, or natural gas, but the process of mining it, processing it, and converting it into energy has an unacceptably high environmental risk.

People against nuclear energy would like to see a future in which the financial assistance given to the nuclear industry is ended and new forms of sustainable energy generation become the norm instead. Developing new technologies in clean, sustainable energy sources is their ultimate goal.

WHAT DO YOU THINK?

Do you think problems associated with nuclear energy, such as storage of spent fuel and waste heat, are reasons to avoid nuclear energy production completely, or do you think they are engineering challenges that can be overcome in order to benefit from the advantages of nuclear energy?

Do you agree that any amount of radiation is unsafe?

Are the threats of terrorism and nuclear proliferation great enough that nuclear energy should not be used in the United States?

Would you feel comfortable living in a town with a nuclear power plant nearby? Why or why not?

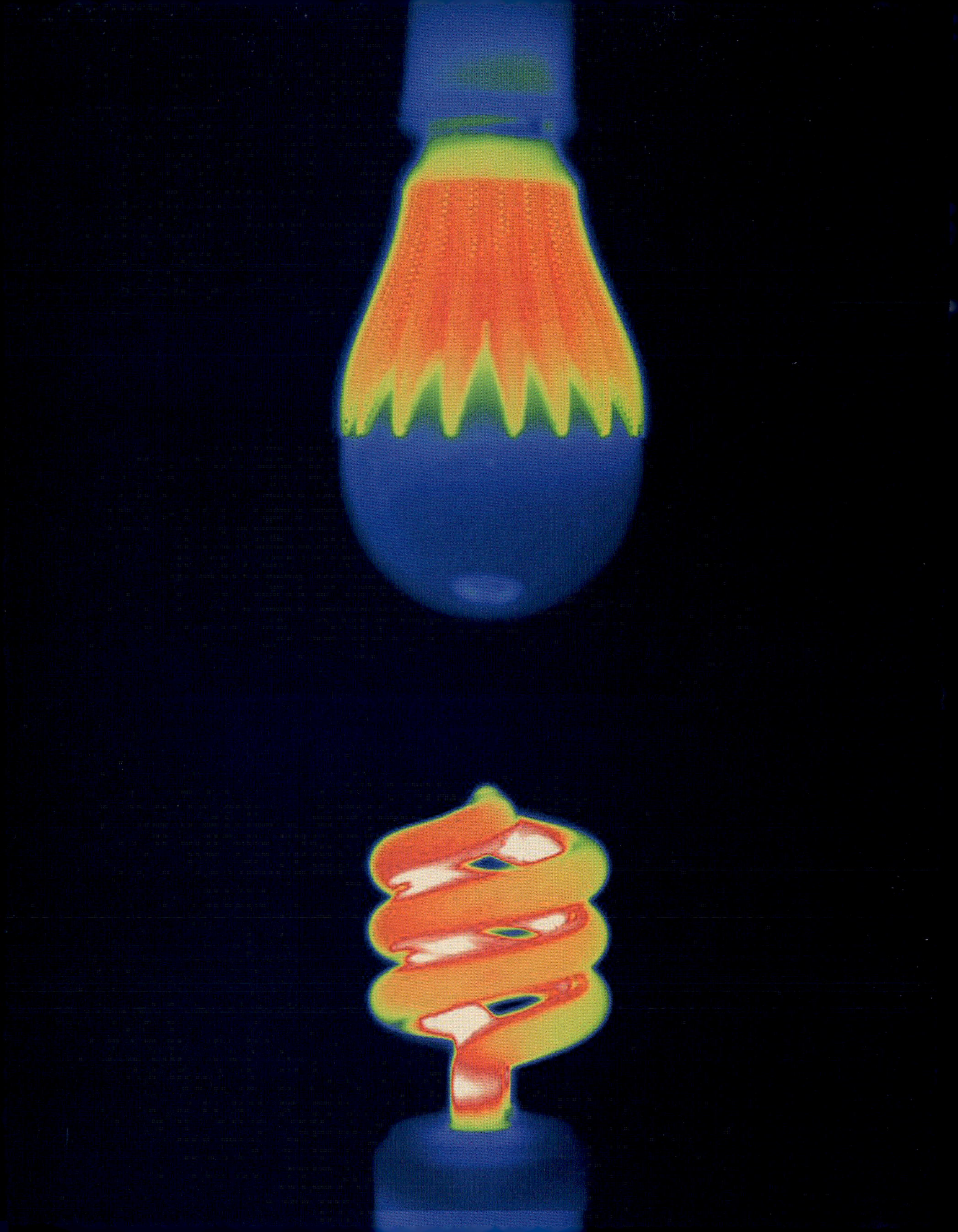

Chapter 4

YOU DECIDE

People in the United States and all over the world are coming to a major turning point in the discussion of energy. They are realizing that fossil fuels are not sustainable as a main source of energy production. Governments and citizens alike understand that new environmentally friendly, cost-effective, and safe energy sources must be sought. Nuclear energy is one of the main topics in this debate. It is poised to become an even bigger issue in years to come.

As of 2011 there were 104 nuclear reactors in the United States. All of them were built more than thirty years ago. These nuclear reactors are now considered old. In order for nuclear energy to be an ongoing option to provide a significant amount of energy in the United States, new nuclear reactors would need to be built. Before any reactors are built, rigorous debate will take place about whether this country should rely on nuclear energy or another energy source. That debate has ramped up. In 2010, President Obama supported $8.3 billion in loan guarantees to help build a brand-new nuclear reactor in Georgia. Companies in South Carolina, Maryland, and

Using more efficient energy is one way we can all help to conserve Earth's natural resources. This photo depicts the efficiency of two advancements in lightbulb technology: LED (top) and compact fluorescent (bottom) lightbulbs.

NUCLEAR ENERGY

Texas are also hoping to win federal loan guarantees to help build new plants in these states.

Most people who feel strongly about nuclear energy have at their core a deep concern for preserving the environment. Those in favor of further development of nuclear energy point to its ability to produce cleaner energy. Those against nuclear energy are concerned about the safety of spent fuel and the catastrophic damage that could occur with future nuclear accidents. The nuclear energy debate is complex. It involves science issues, practical issues (such as where to store nuclear waste), economic considerations, political concerns, and even international terrorism concerns.

As a good citizen, you can be part of the debate. Whether you believe that nuclear energy is a valuable part of reducing reliance on fossil fuels or whether you believe the risks of nuclear energy outweigh the benefits, your voice can color the debate. Writing letters to your

Carmakers are doing their part to help implement more energy-efficient vehicles, such as this plug-in hybrid build by Toyota.

representatives in Congress is one way you can make your voice heard. Participating in formal debates at school or in debate clubs is another way. Even casual conversations among friends and family about the problem of clean, sustainable energy can go a long way in making your voice resonate.

As carbon emissions become a greater problem and climate change continues to threaten the planet, Americans will have to give hard thought to their energy usage. In addition, they will need to think about where the energy they use is coming from.

USING ENERGY RESPONSIBLY

No matter where energy comes from, it is everyone's responsibility to use energy in an efficient way. Here are some ways you can use energy wisely:

- Talk to your family about using energy-efficient materials in your home. Visit www.energystar.gov to learn more about these materials and also to learn if your household qualifies for rebates or tax credits for buying energy-efficient products for your home.

- Set the thermostat to cooler temperatures in winter and warmer temperatures in summer, especially at night and when no one is at home. Recycle as much as you can. If recycling is not offered in your community, think of creative ways to raise awareness, such as displaying art projects made out of recyclable materials.

- Call your energy provider and ask about energy efficiency. Many companies have programs and free offers that help consumers become more aware and use energy more efficiently.

- Turn off the lights when you leave a room. Every little bit helps.

Recycling is something everyone can do to help preserve the earth's resources and live greener lives.

Most people agree that it is in the best interest of the planet to be careful about the use of its resources. People can take little steps each day to reduce energy usage. Using compact fluorescent lightbulbs, turning off lights in empty rooms, and unplugging chargers for cell phones and mp3 players when they are not being used can help conserve electricity at home. Whether the electricity is generated through nuclear power plants, coal plants, solar energy, wind energy, or another method, conserving that electricity is a smart move.

YOU DECIDE

WHAT DO YOU THINK?

After learning more about the pros and cons of using nuclear energy, do you think nuclear energy should be used in the United States? Why or why not?

What do you think are the most serious issues surrounding the nuclear energy debate? How do those issues shape your opinion about the issue?

What are some ways you can become more active in the debate about nuclear energy?

How can you work to reduce energy consumption in your personal life?

What are some ways your school or town could reduce energy use?

Glossary

acid rain—Precipitation whose increased acidity is caused by atmospheric emissions. A rain that can be damaging to plants, aquatic life, and the environment.

atom—A microscopic particle. The building block of all matter.

chain reaction—The process in which neutrons strike atoms and in so doing release more neutrons, which in turn strike more atoms. The reaction is self-sustaining.

climate change—The change in overall weather patterns, of which global warming is a component.

DNA—Protein that is the building block of life; it is susceptible to mutation when exposed to radiation or other environmental toxins.

dry cask storage—A method of storing spent fuel in which, after some time in wet storage, spent fuel is sealed within steel casks and then sealed off again with additional layers of protection.

ecosystem—The localized environment in which a specified set of life flourishes.

fission—The breaking apart of uranium atoms, which occurs when neutrons strike them.

fossil fuels—A source of energy that comes from the earth, such as coal, oil, and natural gas.

global warming—The process encompassing an increase in the temperature of the earth's atmosphere and the rise of its sea levels.

graphite—A mineral used in nuclear reactors to slow the neutrons so that the nuclear reaction is controlled.

greenhouse gases—Carbon dioxide and other gases that are released into the atmosphere in part by burning fossil fuels.

GLOSSARY

isotope—One form of an atom that exists in several forms; isotopes have the same number of protons but a different number of neutrons.

mutation—Permanent change to a gene.

neutrons—Particles in the nucleus of an atom that carry a neutral charge.

nuclear proliferation—The spread or increase in use of nuclear weapons.

nucleus—The center of an atom, which contains protons and neutrons.

particulate radiation—Radiation caused from nuclear reactions.

protons—Particles in the nucleus of an atom that carry a positive charge.

radiation—The form of energy that exists as waves or particles in the air.

radon gas—A dangerous by-product of waste rock that is shown to increase the risk of lung cancer in people with long-term exposure.

smog—A mixture of fog and smoke that can be caused by pollutants in the air.

spent fuel—Nuclear fuel rods that have been removed from a nuclear reactor as waste.

sustainable energy—A form of energy that does not rely on finite natural resources. Examples of sustainable energy include wind and solar power.

uranium—The element used as fuel to create nuclear energy.

waste rock—Rock left over after uranium is mined. This rock contains dangerous radon gas and traces of radioactive material.

wet storage—A method of storing spent fuel in ponds so the water absorbs heat and shields the environment from radiation during the first year or so of storage.

yellowcake—A concentrated form of uranium that has been heavily processed.

Find Out More

Books

McLeish, Ewan. *The Pros and Cons of Nuclear Power.* New York: Rosen Publishing, 2007.

Newton, David E. *Nuclear Power.* New York: Facts on File, 2007.

Olwell, Russell B. *The International Atomic Energy Agency.* New York: Chelsea House, 2008.

Townsend, John. *Using Nuclear Energy.* Mankato, MN: Heinemann Library, 2009.

Websites

ENERGY KIDS

http://tonto.eia.doe.gov/kids/energy.cfm?page=nuclear_home-basics

This website, which is run by the United States Department of Energy, offers fun facts, diagrams, and pictures explaining how nuclear energy works. There is also useful information about different methods of energy generation, including geothermal, wind, solar, oil, coal, and more.

GOING NUCLEAR: HOW ATOMIC SCIENCE POWERS THE WORLD AROUND YOU

http://express.howstuffworks.com/exp-nukes.htm

This website explores the nuclear sciences, including nuclear energy, nuclear weapons, nuclear medicine, and nuclear submarines.

STUDENTS' CORNER: U.S. NUCLEAR REGULATORY COMMISSION

www.nrc.gov/reading-rm/basic-ref/students.html#nuclear_energy

The Nuclear Regulatory Commission's students' page provides an in-depth look at nuclear energy, from the science behind splitting the atom to emergency plans and security for nuclear power plants.

Organizations

ANTI-NUCLEAR ORGANIZATIONS IN THE UNITED STATES

Greenpeace USA—www.greenpeace.org/usa/

Nuclear Control Institute—www.nci.org/

Public Citizen—www.citizen.org/

Union of Concerned Scientists—www.ucsusa.org/

PRO-NUCLEAR ORGANIZATIONS IN THE UNITED STATES

Clean and Safe Energy Coalition—www.cleansafeenergy.org/

Environmentalists for Nuclear Energy—www.ecolo.org/

North American Young Generation in Nuclear—
www.na-ygn.org/index/index.html

US Nuclear Energy Foundation—www.usnuclearenergy.org/

About the Author

Johannah Haney is a freelance writer and editor living in Boston. She has written many books for young learners, as well as magazine articles. She particularly enjoys writing about science topics.

Index

Page numbers in **boldface** are illustrations.

anti-nuclear energy stance, 41–53
Atomic Energy Act and Commission, 21
atoms, 10

benefits of nuclear energy, 28–36, 38

Chernobyl, 18–21, 46
China Syndrome, The (movie), 17
Clamshell Alliance, **40,** 41
clean energy, 9
climate change, 8
coal energy, 29, **29,** 30, 31
coal mining, 27, **30,** 34
control room at Oak Harbor, **34**
cooling pools, **14**
cooling towers, **6, 9,** 13

dirty bombs, 49, **50**
dry cask storage, 15–16

emissions, 29
Energy Policy Act, 25
Energy Reorganization Act, 21, 24
Environmentalists for Nuclear Energy, 38

financial costs, 50
fission, 11–13
fossil fuels, 7
fuel rods, 10–11, **14**
Fukushima, Japan, 20, 23, 42, 48

genetic mutations, **46,** 46–47, 49
global warming, 7, 8, 29
greenhouse gases, 7
Greenpeace, 38

Hiroshima, Japan, 47

isotopes, 10

Low-Level Radioactive Waste Policy Amendments Act, 24

Moore, Patrick, 38
mutations, **46,** 46–47, 49

Nagasaki, Japan, 47
natural gas, 29, 30, 31
nuclear accidents, 16–21, 33–34
nuclear energy, how it works, 10–13

nuclear fallout, 47
Nuclear Non-Proliferation Act, 24
nuclear power plants, 13
nuclear proliferation, 24, 35–36, 51–52
nuclear reactors, 9
Nuclear Regulatory Commission (NRC), 13, 24, 37
Nuclear Waste Policy Act, 24, 37
nuclear weapons, **51,** 51–52

oil rigs, 27

plug-in hybrid vehicle, **56**
pollution, 7, **26,** 29
Price-Anderson Nuclear Industries Indemnity Act, 24–25, 46
Pripyat, Ukraine, **19,** 20–21

radiation, 9, 13, 15, 34, 44
Radiation Exposure Compensation Act, 44
radon gas, 43–44
Rancho Seco nuclear power plant, **32**
renewable energy, 32
risks of nuclear energy, 33–35

safety regulations, 13, 21, 24–25, 48
Seabrook nuclear plant, **40,** 41
shipping nuclear waste, 35
solar panels, **32**
solar power, 32
spent fuel, **14,** 15–16, 24, 31, 32, 34–35, 44–46, 51
Springsteen, Bruce, **52**
storage of spent fuel, **14,** 15–16
sustainable energy, 9

terrorism, 35–36, 48–50
thermal pollution, 43, **43,** 44
Three Mile Island, 16–18, **17,** 33–34
timeline, nuclear energy, 22–23

uranium, 10–12, 30–31, 43–44, 45, **45**
using energy responsibly, 57

waste products. *See* pollution; spent fuel
waste storage sites, 34–35, 37, **37,** 44–46
wet storage, 15
wildlife, 46–47
wind power and turbines, 29, **29,** 32
World War II, 47

yellowcake, 45, **45**
Yucca Mountain Repository, 37, **37,** 44–45